Simulation Series

A Series of Simulations
Designed to Challenge
High-Ability Learners
in History, English,
and the Humanities

Western Explorations

By Charlene Beeler, M.S.Ed.

Prufrock
Press
Post Office Box 8813
Waco, Texas 76714-8813
1-800-998-2208

Table of Contents

Forward

When Charlene Beeler submitted her simulations to me, I was excited. This teacher of almost twenty years had designed a set of four units which were highly appropriate for high-ability learners in the secondary classroom. The units, I felt, could be easily used as models for educators seeking fresh ideas and methods for teaching high-ability learners.

Of course, it should be noted, that the units within the *Simulations Series* offer examples of one strategy for teaching high-ability youngsters—a simulation. There are many others. The series was not designed to offer the strategy for teaching these students—it simply acts as an example of one strategy.

All of the units within the series are designed to be modified by you before they are used in the classroom. These units are not the "anybody-can-teach-em" units I so often encountered as a classroom teacher. These units act as curricular shells which should be modified to align with your personal objectives for your students and the curricular objectives of your district. While preparing for this unit, you will want to add objectives and tasks and develop appropriate evaluation tools.

While you use these units in your classroom, your role will be that of a facilitator or resource person. You will help students locate information and materials. You may also act as the evaluator of student products. However, your importance in this latter role may vary depending on the level of independence your students have gained.

Most importantly, enjoy the units in the series. The excitement and scholarship they promote in the classroom is a real pleasure to observe.

—Joel E. McIntosh
Editor of *The Prufrock Journal*
The Journal of Secondary Gifted Education

Teacher Handbook

Introduction

In this simulation each student in your class will take on the personality of a character from the Old West. Each student will plan a strategy and present his or her views about a hypothetical incident during the simulation of a trial. This simulation provides students with the opportunity to verbally and visually express their opinions and ideas about the American Old West though art, writing, and drama.

I designed this simulation to give high-ability learners the opportunity to explore some of the characters of the Old West in various ways. In this unit, students are given a great deal of variety in terms of the products they develop. For example, just scanning the projects available to students, one can see that students are allowed to develop a creative essay, design an illustration, write a character analysis, and participate in a simulated trial.

As the teacher, you may find that your traditional role changes during this simulation. Your task will be to aid students as they seek the resources they need to conduct their work. In a sense, you will act as a facilitator or resource person for your young scholars.

Products

During the unit, students are asked to develop various products. They participate in a trial, depict a view of the times either through drawings or written descriptions, and write an essay focusing on the life of an individual from the Old West.

Also, the drawings and essays of the students will be collected in a bound anthology, *Our Wild, Wild West*, which may then be shared with students in other classes. This anthology may also be kept for viewing in your room or placed in the library. A cover for this anthology is provided at the end of the Teacher's Handbook, but you may want to have students design a cover of their own.

Structuring Study

Keep the purpose of *Western Explorations* in mind. It was written to provide students with the opportunity to study a number of aspects of various characters of the Old West in an in-depth fashion. It is not the purpose of the simulation to provide a survey of western history during this period.

If your students are not already familiar with the general history of the this period, it would be highly appropriate to take a period of time prior to the simulation to introduce this period.

Of course, you might reverse this order. Some teachers have used this simulation prior to teaching a survey of the period. In this way, students in the class are already "experts" on various characters of the period and can share their knowledge. Both methods have been successfully used by teachers.

Of Time and Grades

The "Unit Time Chart" is certainly not set in stone. The time chart proposed in this handbook was designed for a class of motivated students who were comfortable with research methods and documentation procedures. It is quite reasonable that you might need to adapt the "Unit Time Chart" for students who will need more time.

Also, I have proposed a grading system in this handbook—three project grades, one for the character analysis, one for the illustration or creative essay, and one for participation in the trial. I have also provided a set of criteria for evaluating each of these projects. However, I have found that there is much variation among teachers, classes, and the grades that are awarded. You are encouraged to adapt my suggestions or throw them out and develop your own.

In fact, feel free to experiment with the simulation in general. Make this simulation more than just a one time activity—shape it so that it becomes a part of your "teaching toolbox!"

Student Objectives

Knowledge

1. Identify several characters who were prominent in the history of the Old West
2. Describe the biographical history of a prominent character from the Old West.
3. Identify facts about life in the 1800s in the Old West.
4. Apply knowledge of events and characters to a simulated situation.

Skills

1. Organize research data into a number of formats.
2. Communicate information through oral discussion, written presentation, and artistic media.
3. Analyze past events.
4. Use reference sources to gather information.

Attitudes

1. An appreciation for other people's viewpoints.
2. Respect for people who lived in another age and the problems with which they dealt: prejudice, avarice, greed, ignorance, and ambition.

Materials

1. Reference materials on the Old West from the library.
2. A copy of the "Student Handbook" for each student (duplicate the "Secret Information" briefings only for those who are to receive the various briefings from it).
3. Drawing paper and paints, markers, and/or colored pencils.
4. A gavel.
5. A dark robe for the judge.

Initial Teacher Instructions

1. After reading through the "Teacher Handbook" and the "Student Handbook," assign to each of your students a role listed on the characters page. All characters with an asterisk by their name should be assigned first—these are the individuals who have specifically assigned duties during the trial.

2. For each student in your classroom, make a copy of the "Student Handbook."

3. Make a single copy of the "Secret Information" section.

4. Obtain and organize reference materials for students to research their characters or instruct them to search in the library.

5. Provide visual material related to the Old West for display and motivational ideas.

Instructions for Play

1. Gather students to the front of the room in a tight circle.

2. Hand out the "Student Handbooks."

3. Read the "Overview" together.

4. Indicate to your students any visual materials you have available and discuss what students already know about the Old West: values, landscapes, people, occupations, and system of justice.

5. Brainstorm words that they consider appropriate for that time: gunfights, saloons, horses, carriages, Native Americans, etc.

6. Explain that each student will receive a character to play during a simulated murder trial. It is their responsibility to research and take on the personality of those people as if they were preparing for parts in a movie.

7. Call the students' attention to the "Character Role Descriptions" pages in the "Student Handbook," and ask students to read through the various parts discussed.

8. Assign the various characters to the students in your class. Urge students to "get into the part" and embellish whenever possible.

10. Ask your students to turn to the section titled "Character History." Explain that this is their first assignment to complete. Discuss with students the "Character History" instructions. If students are unfamiliar with the idea of internal documentation or a bibliography, you should plan to incorporate a lesson on this topic at this time. Instruct students that while all information they provide in their profiles must be historically accurate, they may need to extrapolate some information from historical fact. For example, one student might wish to describe the character's favorite food. More than likely, such specific information will not be available in historical documents; however, that student will certainly be able to identify foods generally eaten during the period and choose from those. Note that Blue Duck will not have to complete this assignment. The student playing this part should work on either the illustration or creative essay described below.

Then, while the rest of the class works on these latter projects, Blue Duck can help with his own defense.

11. Turn to "The Situation." Ask a student to read this orally while the rest of the class reads silently.

12. Explain that after they have completed the "Character History" assignment, they will either begin preparing for the trial (if they are playing one of the attorneys or Blue Duck) or they will develop an illustration or creative essay related to the period.

13. Review the "Illustrations or Creative Essay" section with your students.

14. Finally review the "Simulation Participation" form with your students. This will give them a good idea of what a "good" job during the simulation constitutes. Explain that some students not directly participating in the trial will act as evaluators by filling out the "Simulation Participation" form for each student participating in the simulation. At this point, offer any special instruction concerning this matter.

Unit Time Chart

Phase 1 (Approximately 3 Days)

1. Introduction to the simulation.
2. Research character roles.
3. Begin work on Character Histories.

Phase 2 (Approximately 5 Days)

1. Trial preparation.
2. Begin work on illustrations and creative essays (for all non attorneys).
3. Completion of Character Histories.

Phase 3 (Approximately 2 Days)

1. Trial.
2. Completion of illustrations and creative essays.

Phase 4 (Approximately 1 Day)

Debriefing (includes evaluation of the trial, thought questions, and evaluation of simulation play).

Debriefing Questions

1. What was the most important fact you learned?

2. What were some of the most interesting facts that you learned?

3. What do you feel was the best part of the trial?

4. How did you embellish your part in the trial?

5. Did you agree with the jury's decision?

6. Is there a difference between the old way of administering justice and today's? How is it different?

7. Would you like to have lived in the Old West? Why or why not?

8. What do you imagine life for teenagers was like during the time?

9. Who was the most interesting character you encountered during the simulation?

10. What did you like best about this simulation?

11. Would you like to participate in another simulation?

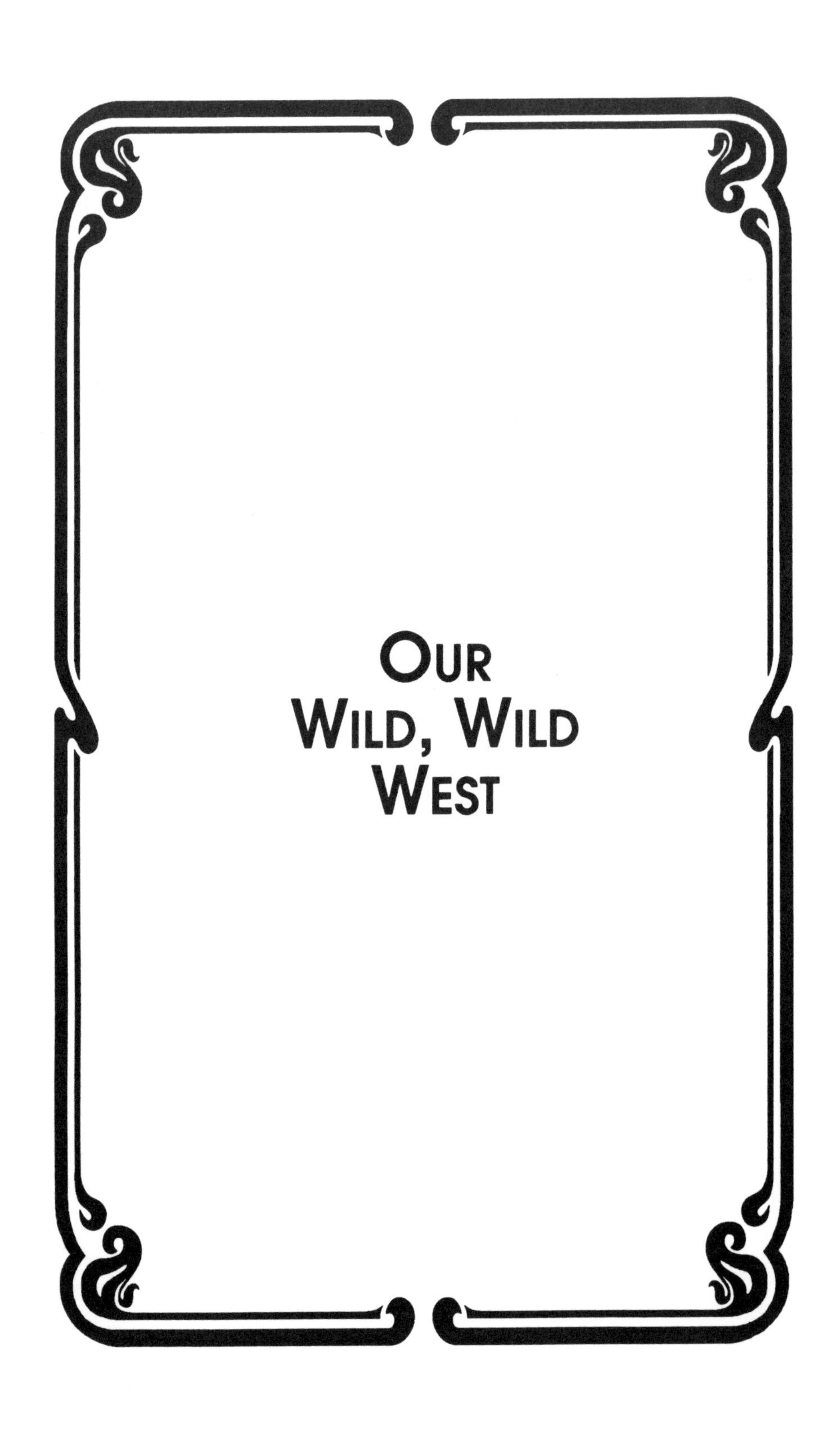

Our Wild, Wild West

Student Handbook

Overview

In the Old West, Native Americans sometimes provided outlaws with predatory opportunities and a cloak for their deeds. In 1836, for instance, a punitive expedition was launched against a Native American "war" party rumored to be a thousand strong which turned out to an indignant but conspicuously peaceful little band. Native Americans were often accused of committing crimes they did not commit. Yet, it certainly cannot be said that Native Americans never committed crimes during the period.

In this simulation, you will participate in a murder trial from 1836. A white man, Blue Duck, who befriended and traveled with Native Americans has been accused of the murder of Will Green. Is the accusation of murder in this trial the result of prejudice against the Native Americans with which Blue Duck travels, or is the accusation true—is Blue Duck a murderer? As characters of the time, you will participate in a trial to discover the truth, if you can, about Blue Duck. Did he commit a crime, or was Blue Duck a scapegoat merely because he was in the wrong place at the wrong time? These questions will be answered at the conclusion of the trial—by you.

We will play with history a little in this simulation. While the trial actually occurred, we have changed some of the characters involved. In fact, we have brought to this simulated trial both heroes and villains from across the Old West. You will take on the personality of one of these characters. Think, act, and react as this person would. Have fun. Experience the emotions, the ideas, and the feelings of your character. Play Doc Holliday, Judge Roy Bean, or Sitting Bull!

During this unit, you will participate in three projects. You will develop a creative character analysis, you will draw an illustration of a scene from the Old West or write a creative essay, and you will participate in the trial of Blue Duck. You will receive three grades during the unit: one for your analysis, one for your illustration or creative essay, and one for your participation in the trial.

Unit Time Chart

Phase 1 (Approximately 3 Days)

1. Introduction to the simulation
2. Research character roles
3. Begin work on "Character History" assignment.

Phase 2 (Approximately 5 Days)

1. Trial preparation.
2. Begin work on illustrations and creative essays (for all non attorneys).
3. Completion of "Character History" assignment.

Phase 3 (Approximately 2 Days)

1. Trial
2. Completion of illustrations and creative essays

Phase 4 (Approximately 1 Day)

Debriefing (includes evaluation of the trial and evaluation of simulation play).

Character Role Descriptions

Become familiar with all of the following people by reading the briefs from these pages. You will most assuredly come in contact with each and every one of these folks during the trial. By knowing your enemy or your friend, you will be able to conduct yourself with forethought, cleverness, and knowledge.

1. **William Clark Quantrill***—a former Bible teacher turned murderous villain. His sympathies lay with the confederacy, but his notions of warfare were more in common with Attila the Hun than with Robert E. Lee. He wore a hatchet, a saber, and as many as 8 pistols in his belt.

2. **Wyatt Earp***—did not become a deputy marshal to tame the town as he claimed. He first began wearing the garb of a professional gambler and lived by his wits. He then became a bill collector and a policeman. He was involved in a dispute over cattle and dismissed from the police force. Later he went to Dodge City and claimed to be a marshal.

3. **Judge Roy Bean***—wore a sombrero and a six-gun. Judge Bean was the law of the land. He metered out law harshly.

4. **William Bonny (Billy the Kid)***—outlaw. He was 17 years old when he killed his first man because the stranger was making time with his girl friend. Later, he became a notorious gang leader, robbed banks, and killed many people. "The Kid" became an American legend.

5. **Sheriff Pat Garrett***—a former gang member of "the Kid." After turning from his life of crime, he was hired to trap or kill Billy the Kid. It was said he shot the outlaw in the back and called it justifiable homicide. Nonetheless he lost the next election for his sheriff's position, and he spent his remaining years as a not so prosperous rancher.

6. **Tom Horn***—braided the rope for his own hanging. He had been hired by a group of cattle ranchers to track down trouble makers (woolies—sheep ranchers) who were giving cattle barons a hard time. It was believed that sheep overgrazed, tramped the land and muddied up the watering holes so that cattle couldn't drink. When Horn was arrested for murdering a 14-year-old boy, he explained

that the killing was accidental. The cattlemen never came forward and helped him.

7. **Belle Starr***—a woman of easy virtue and riotous living. Ranching, rustling, and horse trading gave her a good living. She was also good at bribing juries and enjoyed being called a jury "fixer." She was said to prefer the company of Native Americans and outlaws. She was fond of Blue Duck, a horse thief. She later married Cole Younger.

8. **Blue Duck***—a white man who befriended and traveled with Native Americans. Blue Duck is a horse thief and Belle Starr's lover.

9. **Cole Younger***—an American desperado.

10. **Allan Pinkerton***—formed his own detective agency. Was hired to hunt down Butch Cassidy and the Sundance Kid, among others, and to protect railroads from robbers.

11. **Red Cloud**—a leading warrior of the Sioux nation who was savvy enough to bargain with the white man and gain a reservation for his tribe that today spans most of South Dakota.

12. **Chief Joseph***—a brilliant and cunning Native American who led the army on a wild goose chase all over the west in the 1870s.

13. **Sitting Bull***—Dakota chief who, along with Big Foot, another Native American leader, led the uprising known as the Battle of Wounded Knee.

14. **Geronimo***—a powerful Apache leader often feared by white men.

15. **Col. George Armstrong Custer***—wanted to be famous soldier. He was successful, although reckless in his attempts and ambitions. He had been a Civil War soldier and fought well. His bravery earned him a command. Custer met his end at the Battle of Little Big Horn.

16. **Jedediah Smith***—an unusual frontiersmen—he neither drank alcohol nor smoked tobacco.

17. **Jim Bridger***—a frontiersmen, fur trader, and guide.

18. **Christopher (Kit) Carson**—Scout, fur trapper, and Indian agent. He led a life from which legends were made. He was a popular hero among many of the time.

19. **James Butler (Wild Bill Hickok)***—a corrupt lawman who dressed for his public, wore his hair and moustache long, and might have been handsome but for his prominent nose and lips.

20. **Emmet Dalton***—a member of the Dalton brothers, a notorious gang.

21. **Bob Dalton**—a member of the Dalton brothers, a notorious gang.

22. **Bill Dalton**—a member of the Dalton brothers, a notorious gang.

23. **Grant Dalton**—a member of the Dalton brothers, a notorious gang.

24. **Bill Doolin**—a notorious bank robber.

25. **The Texas Rangers***—made a name for themselves as brave men who tried to keep the west from chaos. They fought bank robbers and cattle rustlers and protected white settlements against Native American raids.

26. **Quanna Parker***—white woman who was kidnapped by a Native American tribe and lived most of her life with her captives. She married the tribe's chief and had a son. Later, when rescued by her original people, she had trouble re-adjusting to her new life.

27. **John Henry (Doc) Holliday***—a western frontiersmen.

Role Assignments

1. Defendant—Blue Duck
2. Judge—Judge Roy Bean
3. Prosecuting Attorney—Col. George Armstrong Custer
4. Attorney for the Defense—Wyatt Earp
5. Bailiff—Doc Holliday

Witnesses for the Prosecution

6. Tom Horn
7. James Butler
8. William Clark Quantrill
9. William Bonny

Witness for the Defense

10. Belle Starr
11. Jim Bridger
12. Pat Garrett
13. Chief Joseph

Jury

14. A Texas Ranger
15. Jedediah Smith
16. Sitting Bull
17. Geronimo
18. Quanna Parker
19. Emmet Dalton
20. Cole Younger
21. Surprise Witness—Allan Pinkerton

Participation Evaluators

All additional characters will act as evaluators during the trial. Students acting as evaluators will use the "Simulation Participation" form to evaluate each student participating in the trial. Your teacher will provide any necessary special instructions related to this role.

Character History Project

You will develop a personality profile. This profile will take the form of an informal essay focusing on the life of your character. The essay should be in first person and tell the story of your character's life. You should write as if the character were writing for an audience of modern teenagers. Include information about your character's personality, major accomplishments, intriguing experiences, and significant relationships. Include any information you feel is important or interesting.

- Each paper must provide researched evidence. This evidence should be internally documented (your teacher will describe internal documentation if you are unfamiliar with it). Do not simply copy from a reference book, integrate your evidence within the first person narrative of your paper.
- The paper should refer to at least *five sources of information.*
- Be sure to include many specific historical details.
- You should include a Works Cited section at the conclusion of your paper.

The style and content of your essay will be evaluated using the following criteria and others which your teacher will describe:

- Was the essay written in first person from the perspective of the student's character and for an audience of teenagers?
- Did the student include at least five sources of information in this project?
- To what extent did the student include a variety of historically accurate details about the character?
- To what extent was the author's work clearly directed toward a audience of teenagers?

Of course, be sure that your final draft is free of grammatical errors. Your essay along with others from the class will be included in a classroom booklet to be used and read by other students in other classes.

Illustration or Creative Essay Project

During this portion of the unit, you will choose either to illustrate a scene from the Old West or to write a creative essay examining modern issues through the eyes of the character you have been assigned. Note that the attorneys are not required to develop this project. Regardless of the project you choose, you should keep the following criteria for evaluation in mind:

- Your illustration or essay should contain at least *five historically accurate representations*. For example, in task one, you are asked to draw a street scene from a typical western town. Your drawing might include historically accurate clothing, foods, weapons, building design, and speech (in cartoon captions—perhaps a copy of *The Adventures of Huckleberry Finn* by Mark Twain would be helpful in that it offers examples of many dialects from the period). Or, for instance, if you are writing a creative essay, you might have your character note the differences between school discipline in modern times and school discipline in the 1800s.
- Document the source of your information in determining that a representation is historically accurate by listing each representation and the complete bibliographic information of its source on the back of your drawing or internally documenting your sources in your creative essay. For example, a reference on the back of an illustration might be similar to the following: "All Native American clothing in the picture is described on page 190 of Josephy, Alvin. *The Patriot Chiefs*. New York: Viking."
- The complexity of the work is very important. Include many historically accurate details in either your illustration or your essay.

Topics of Illustrations

- A street scene from a typical western town
- Outlaw portraiture
- Gun fights
- Range wars
- The gallows
- Landscapes
- Border clashes
- Native American farming or hunting scenes
- Native American pottery or jewelry design
- Animals of the plains or mountains
- Native American villages
- Native American art

Creative Essay Topic

Imagine that the character you will play in this simulation has been hurled through time and space into another century—the late 1900s. Of course there will be a culture shock with which you will have to contend. What are your emotions, feelings, and perceptions of this new time?

Write an essay describing what you see. Comment on the parallels between this modern age and the age from which you came. What are the differences? What changes do you find most shocking? Be sure to describe the things in the 1800s which were similar to or different from the things you see in the 1900s.

The Situation

About noon, on June 28, 1888, a group of Blackfoot tribe members stumbled into town from their nearby camp, half drunk and rowdy from the previous night's liquor binge. It was unclear how they got ahold of the liquor, but the result was always the same. Trouble.

At first there was a lot of merriment, some recklessness, and a modicum of disruption and irresponsibility among the braves. They were running and shooting their guns. Some stumbled into the saloon to drink some more. The same thing happened when cowboys, heated and ready for action, rode in from a cattle drive. Loud, rude talk and boasting was a large part of the activity. Most of the time, the townsfolk ignored them or got out of their way the best way they could. Merchants did not want to upset any of these men because their money was valued. Of course, there was always somebody who didn't want to ignore the arrival of what they considered "out of place nuisances" and made a big deal out of it, threatening to take care of "them."

Noon arrived with some floury. Two shots rang out and reverberated through the streets of town. Everyone turned in the direction of the saloon where the shots originated. The doors banged back against the outside wall. Blue Duck, a white man of unknown origin, who was known for running with the Native Americans, was being carried out of the "Greenhorn Saloon." Several men dumped him on the street and readied to shoot him in cold blood. It seemed that Will Green, the proprietor and owner of the saloon, had been shot, and Blue Duck was accused of killing him. He was found with a gun in his hand.

Blue Duck was dazed and confused. A mob formed around him. But, after a few harrowing moments, a man with a beard and long hair tied in a pony tail at his neck, shoved his way through he crowd. Jedediah Smith, a mountain man, spoke over the din of the crowd. "Hear me out. There will be no more killing today. Blue Duck deserves a trial."

And so it was. Blue Duck was put under armed guard and held for trial. The townspeople eagerly awaited the event as if it were a holiday. Gossips hashed and rehashed the incident as if they were on the scene and had seen exactly what had happened. The truth was, no one was really sure what had transpired. The confusion surrounding the incident was so intense, and the people involved were so ... well, let's say, interested in taking care of themselves, it was hard to get at the real truth.

But of course, that will be your job. Sifting through testimony and making sure real justice is done will be your responsibility. What happens next is up to you—good luck.

General Trial Procedures

1. The bailiff will ask everyone in the courtroom to rise, and will announce the judge.
2. After the bailiff has announced him, the judge will tell those in the courtroom to "be seated."
3. The judge will instruct the jury members to listen to all testimony before making up their minds.
4. The judge will begin the trial. He or she will instruct the attorneys to make their opening statements (the prosecuting attorney will speak first, then the defense attorney will speak).
5. After the opening remarks have been given, the judge will instruct the prosecuting attorney to call his or her first witness.
6. The witness will be sworn in by the bailiff and questioned by the prosecuting attorney.
7. The defense attorney will then cross examine the witness.
8. Upon conclusion of number seven, the next witness on the list will be called. Steps six and seven will be repeated until all witnesses for the prosecution have been called.
9. Next, the judge will ask the bailiff to swear in the first witness for the defense and then allow the defense attorney to question the witness.
10. The prosecuting attorney will then cross examine the witness.
11. Upon conclusion of number ten, the next witness on the list will be called. Steps nine and ten will be repeated with the remaining witnesses until all witnesses for the defense have been called.
12. The prosecuting attorney will cross examine the witness.
13. At the conclusion of testimony, the attorneys will provide closing statements.
14. The judge will instruct the jury to adjourn and deliberate its verdict. The other members of court will be dismissed to work on their projects.
15. When the jury returns, the judge will ask the foreman of the jury to give the jury's verdict.
16. The judge may poll the jury for individual votes if he or she wishes.

Simulation Roles and Duties

Judge

Materials

1. A gavel of some sort
2. A desk or table
3. A dark robe

Duties

Without doubt, your job is the most important. Keep the trial running smoothly and efficiently by following the procedures below.

1. Wait for the bailiff to announce you. After the bailiff has finished his speech, tell the people in the court to "be seated."
2. Instruct the jury members to listen to all testimony before making up their minds.
3. Begin the trial. Instruct attorneys to make their opening statements (the prosecuting attorney will speak first).
4. After the opening remarks have been made, instruct the defense attorney to provide you with a list of witnesses.
5. Instruct the prosecuting attorney to call his or her first witness.
6. Ask the bailiff to swear in the witness and then allow the prosecuting attorney to question the witness.
7. The defense attorney will then cross examine the witness.
8. Upon conclusion of number seven, the next witness on the list will be called. Repeat steps six and seven until all witnesses for the prosecution have been called.
9. Hear any objections from the attorneys. Overrule or sustain their motions. Essentially, because the rules of court room procedures were applied less rigidly in the Old West, your job here is to determine the essential fairness of the objection or motion being made. If an attorney begins objecting too often, simply threaten to refuse to entertain any objections from that lawyer and follow through if necessary.
10. Defense attorney will provide a witness list and call the first defense witness.
11. Ask the bailiff to swear in the witness and then allow the defense attorney to question the witness.
12. The prosecuting attorney will then cross examine the witness.

13. Upon conclusion of number eleven, the next witness on the list will be called. Repeat steps eleven and twelve until all witnesses for the defense have been called.
14. At the conclusion of testimony, instruct the attorneys to provide closing statements.
15. Instruct the jury to adjourn and deliberate its verdict. Give them a time limit (twenty minutes or so). Dismiss other members of court to work on their projects.
16. When the jury returns, ask the foreman of the jury to give the verdict.
17. You may poll the jury for individual votes, if you wish.

Attorneys

Materials

1. A table
2. A trial brief prepared by you

Duties

1. Carefully read over the information provided for you in the Characters pages. Become familiar with the witnesses, your jury, the judge, the defendant, and everyone connected with this trial. Anticipate your opponent's moves by knowing as much as possible.
2. Interview your witnesses carefully before the trial (deposition). Ask such questions as:
   a. What is your relationship to the accused?
   b. How long have you known him or her?
   c. Where were you on the date in question?
   d. What did you see and hear?
   e. What is your opinion of the accused?
   f. Do you know anything else, incriminating or helpful?
3. Make a list of witnesses in the order you wish to call them during the trial and give a copy of your list to the judge and the opposing attorney.
4. If you wish, you may conduct brief (five minute) depositions (questioning the opposition's witnesses); however, you must do this in the presence of the opposition attorneys.
5. It is up to the defense attorney whether or not to call upon the accused to testify.
6. Write an opening statement (no longer than five minutes) that briefly tells the jury and the judge what you propose to do. For example, "I propose to prove to the court that my client was not responsible for the death of Will Green because he has never shot an unarmed man, and character witnesses will tell you that he is not capable of such unfair violence ..." or "I propose to show you that my client acted in self defense after being ridiculed and threatened with his life ..."
8. The prosecuting attorney might say something like. "I will attempt to prove that the defendant did willfully murder Will Green and had contemplated the action previously ..."
9. Attorneys may object to witnesses' statements or the other attorney badgering a witness; however, the judge will be ruling on the essential fairness of the objections you raise, so be sure your objections are valid.

10. During the trial, you will be given the chance to cross examine the witness, so prepare for these situations.
11. The judge will ask for closing statements at the conclusion of testimony, so be prepared to deliver such a statement.
12. After closing testimony, your job is complete except for listening to the verdict and consoling or congratulating your client.

Note that as the attorneys, you will not be required to complete the illustration or creative essay project of this unit. While your classmates work on this project, you will prepare for the trial.

Witnesses

You will be questioned by both attorneys. Tell them as much as you deem fit. Read the character's roles so that you will be prepared to give a good or bad character reference. Act, react, and pay attention. The simulation and the outcome of the trial depend on you.

Make absolutely sure that what you say is consistent with your character and follow the rules below.

- With one exception, you may *not* lie.
- The exception is that you may lie in cases where telling the truth would directly incriminate you.
- You may attempt to avoid answering questions through redirection; however, avoid using this tactic too often and answer a question directly when told to do so by the judge.

Jury

Listen very carefully to the trial testimony. Take notes. Watch expressions. Read very carefully the character role descriptions. Be careful—some witnesses may lie, but witnesses are allowed to lie only when telling the truth might incriminate them.

Be prepared to argue your opinion with other jurors near the trial's conclusion. Are you persuasive? The success of the simulation and the trial will depend largely on your decisions.

Defendant

Act, react, and play on the jury's sympathy. Read the character role descriptions very carefully. Know about the witnesses who will testify against you. Do they have ulterior motives to rid this planet of you? Do you have a suspect? Remember you are playing the role of Blue Duck, not yourself.

Note that as Blue Duck, you will not be required to complete the "Character History" assignment. While the other students are working on this assignment, your should complete the illustration or creative essay project of this unit. Later, while your classmates work on this project, you will help your attorney prepare for the trial.

Bailiff

Duties

1. Announce the entrance of the judge.
2. Instruct the courtroom to stand when the judge enters by reading the following:

   "All rise. Hear ye. Hear ye. The court of the honorable Judge Roy Bean is now in session."

3. Each time a new witness is brought before the court, you will be asked to swear in the witness. Do so by saying the following:

   "Raise your right hand. Do you solemnly swear to tell the truth, the whole truth, and nothing but the truth?"

3. Keep order in the court
4. Announce when court is adjourned.

Evaluation of Simulation Participation

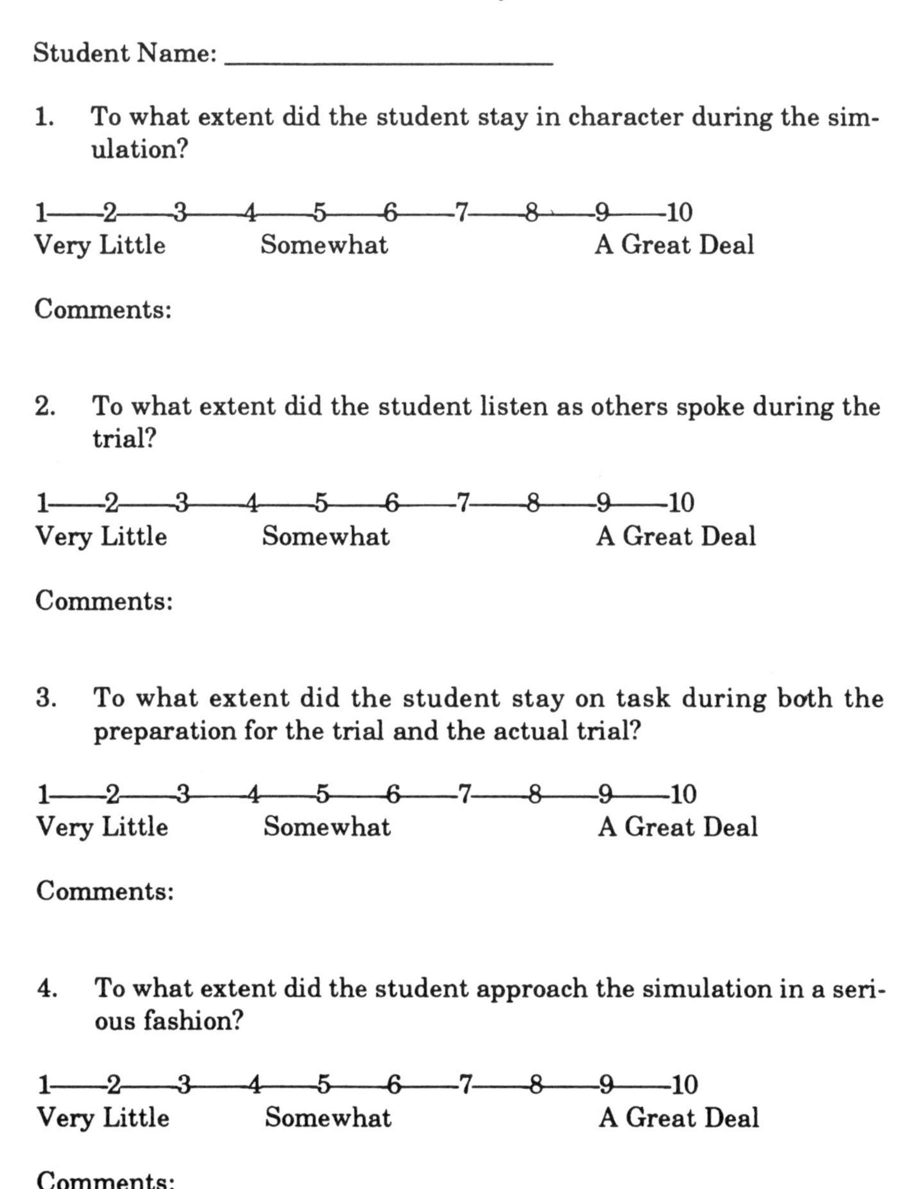

Student Name: ________________________

1. To what extent did the student stay in character during the simulation?

1——2——3——4——5——6——7——8——9——10
Very Little Somewhat A Great Deal

Comments:

2. To what extent did the student listen as others spoke during the trial?

1——2——3——4——5——6——7——8——9——10
Very Little Somewhat A Great Deal

Comments:

3. To what extent did the student stay on task during both the preparation for the trial and the actual trial?

1——2——3——4——5——6——7——8——9——10
Very Little Somewhat A Great Deal

Comments:

4. To what extent did the student approach the simulation in a serious fashion?

1——2——3——4——5——6——7——8——9——10
Very Little Somewhat A Great Deal

Comments:

Secret Information

Secret Information
Blue Duck—Defendant

You were too drunk to really know if you shot Will Green or not, but knowing yourself as you do, you could have done it. You had a six shooter and four bullets had been fired from the chamber. You know you did some practice shooting yesterday, and you don't remember how many bullets you used. It is only too bad that no one thought to check your weapon right after the shooting to see if it had been fired.

There are a few things you do remember. Pat Garret was your partner at cards and cheating at cards. The two of you were cheating at a card game, but you were having a tough time keeping your train of thought. Pat was loosing his patience. His mustache twitched. His hand tightened around his whisky glass. For a minute you thought he was going to accuse you of cheating and shoot you. Pat can be dangerous when he is mad.

Then you saw Will Green striding toward the table staring at you with fire in his eyes. Was it jealousy or greed that turned his face a funny green color? He shouted, "What are you doing in my saloon, you no-good, lily livered heap of cow dung! I don't take kindly to the likes of you and your kind!"

You stood up and tried to focus your eyes. Will Green's ugly sneering expression, his big hands, even his stinking odor permeated through your drunken haze and paralyzed you with fear. This was his territory, and this was going to be the "show down."

If anyone asked you why he was mad, you'd say Belle Starr was the reason. She preferred you instead of him. If he couldn't win her heart fairly, he was going to bully you out of his town ... or shoot you. His reputation preceded him. Will Green was known for playing dirty and low down. He was probably also mad about the bar tab you'd accumulated over the afternoon—about thirty dollars.

But, you wouldn't have any part of being bullied. You wanted to stand your ground. You felt a mixture of adrenaline, fear, and macho pride. You could take on this creep, but your legs felt like rubber. Your skin crawled as though reacting from a shivering attack. He was shouting obscenities. You reached for your gun. You couldn't allow anyone to embarrass you like this. But your hand was shaking so badly you had trouble holding your gun. You tried to raise it and shoot. You then heard two shots coming from different directions—perhaps not from your gun ... then all went black. The next thing you remember was several faces looking down at you, rough hands grabbing your body, and loud voices yelling, "You murderer!"

"Who me?" you asked.

Secret Information
Roy Bean—Judge

You have presided over some of the best and most interesting trials. You are the most colorful judge in the whole land. You have a reputation to withhold. Your sombrero is as famous as your sometimes bizarre sentencing. Do you remember the time you had a bear in your court? No one can accuse you of being dull or a wimp. You look forward to all your trials because they make you feel like a king. After all, you do have your pride, and whether the culprit is guilty or innocent, it is your duty to put the fear of God into each and every one. Your motto is, "Hang 'em high ... if they deserve it and sometimes even if they don't."

Secret Information
Tom Horn—Witness for the Prosecution

You were playing cards on the date in question. You have nothing against Blue Duck. In fact, you and he have ridden together on a few cattle drives—legal and illegal—but you have a secret of your own.

If Blue Duck swings for this killing you will have gotten away with your little crime. After all, Blue Duck can't put up too much of an argument. He was too drunk to remember what he was doing. You could see it in his eyes the moment he came to from his drunken stupor. Play it smart and you'll get away with it—with killing Will Green.

You were contemplating ambushing Green and blaming it on the sheep ranchers. Green had it coming. He was an obnoxious, a little man who liked to hurt people. He was a crooked cattle rancher, and the Cattle Association had offered you a handsome price to get him out of the way—and they didn't care how it was done.

When Blue Duck went for his gun, you made sure you went for yours at the same time. It was obvious Blue Duck wasn't going to pull the trigger, or if he did, he wasn't going to hit anything. He wasn't a killer, just a horse thief. He didn't have the guts to kill people—like you do.

Don't incriminate yourself. You may want to tell the jury how conniving Blue Duck is, how no one trusts him and how he wanted to get rid of Green because of Belle Starr. His motive for killing Green was to rid himself forever of any interference to his romance. After all, everyone knew Belle Starr was fickle. True, today she likes Blue Duck, but she is partners with Will Green in the cattle business.

Secret Information
Jim Bridger—Witness for the Defense

You were playing cards with Blue Duck on the afternoon in question. You've known Blue Duck for several years. He's just a cowboy who mostly minds his own business—when he's not rustlin' cattle. He can't shoot very well either. On a good day, he can't hit the side of a barn. And as far as you know, Blue Duck doesn't have a motive for killing Will Green. Even though Green and Belle Starr were business partners, Blue Duck was not a jealous man—Green was.

You don't really poke your nose into other people's affairs, preferring to allow folks to go about their own way. Live and let live is your motto.

Actually, you didn't see Will Green approach the table. You saw Blue Duck look up and retort something in return to Green's rhetoric. You saw Blue Duck go for his gun and fumble with it. Then, you heard two shots that seemed to come from different locations in the bar (neither of which was where Blue Duck stood), but you really can't be sure, there was a lot of confusion. If you had to guess who would have shot Green, your guess would have been Tom Horn. He's a sly one, ambushin' people in the name of the Cattle Ranchers Association.

Another suspect would be James Butler. You just can't abide a self righteous man who acts like an upstanding citizen when all along he is a side winder with an ace up his sleeve and a gun not far behind.

Or maybe it was Pat Garret. You know, who can you trust around a town filled with outlaws and robbers?

Secrete Information
James Butler—Witness for the Prosecution

All you could think of on the date in question was how you were going to show up Pat Garret as a cheater. You saw right from the start how he made signals to Blue Duck, and how Blue Duck, too drunk to acknowledge the clues, just winked and played any old card he had in his hand. They were loosing, but Garret was smart He would find a way to get rid of Blue Duck and cheat his way to a run of winning hands. And he was slick, slick, slick.

The only way to show up Pat Garret and not get shot in the back at a later date was to trick him. How to do just that was what was on your mind when Will Green stomped onto the scene. All of a sudden, you saw your chance. There was sure to be a show down, a big brawl, or an out-right killing. You waited, your hand on your gun, a smile on your face.

Everybody stood up. People pushed and shouted. Garret was in the thick of things, right next to Will Green, trying to calm him down. He was a prime target. When Blue Duck went for his gun, you went for yours. Did anyone see? You don't think so. More important, did anyone see what happened after that? In actuality, you'd be embarrassed if anyone saw your failed, poor attempt at murder. After all, you have a reputation to uphold. Your bullet missing Garret and hitting Will Green was a bad mistake. If anyone did see, you'd be the laughin' stock of the whole town.

But, with Blue Duck sitting in the defendant's chair, his glazed eyes peering out from his disheveled hair on his tired cowboy's face, the jury will surely think he's guilty. You can't wait for your chance to cement Blue Duck's fate. Getting this whole mess cleaned up and forgotten is your prime goal now.

Secret Information
William Clark Quantrill—Witness for the Prosecution

You've murdered plenty of people in your day. What's the fuss about one more? Your motto is, "If they die, they deserved it, or they'd have been smarter." You're still alive, aren't you?

That's too boring. Besides, you really don't know for sure, but you suspect that Blue Duck is being framed by the real murderer, whoever that is. Maybe that sly Belle Star found a way to get rid of Blue Duck. She's a fickle woman, first wanting this man, then wanting that one. All she had to do was whisper in Will Green's ear that Blue Duck had taken advantage of her, and Green would set out to avenge her honor. What honor? Belle Starr? Who are you kidding?

You've only seen Blue Duck around town, you've never talked to him or seen him conduct business or do any of the things the townspeople say he does. You were in the bar the afternoon in question. You were drinking and thinking about sitting in on the game of cards going on when you saw Green approach the table. You hung around hoping to witness a killing. After all, killing is your line of business. Killing is your game. Yet, in the confusion of the situation, you didn't get the chance to see much of the shooting. In fact, and this is a little strange, the shots you heard didn't sound like they were coming from Blue Duck's direction.

Secret Information
Chief Joseph—Witness for the Defense

You've known Blue Duck for several years.

Blue Duck is a lame brain. You never had any use for him, but he doesn't deserve to hang at the white man's hands. He drinks too much, he steals horses, and he is stupid. Why can't he make up his mind about which culture he belongs to?

But for whatever reason Blue Duck thinks he's a Native American, and you will stand up for him in this trial. After all, the white man takes advantage of the red man all the time and this might as well be a trial of whites against reds.

You did not see what happened in the saloon that day (it happened too fast), but you believe Blue Duck was too drunk to use his gun. You did hear the two shots that were fired, but in your opinion they couldn't have come from the same gun—the shots came much too fast to be fired by a single gunman.

Secret Information
William Bonny (Billy the Kid)—Witness for the Prosecution

You don't know whether Blue Duck is innocent or not, but if Pat Garret is on his side, you aren't. It's as simple as that. So, whatever Pat Garret says, you will refute. Blue Duck is a little man anyway, not worth living. He wouldn't join your gang 'cause he said he was a loner. He's just a cheap horse thief, nothin' big. Why is Pat Garret so generous, anyway? It 's not like him to stand up for a man like Blue Duck. He must have an ace up his sleeve. Watch out for that Garret. You don't trust him.

Secret Information
Pat Garret—Witness for the Defense

It's highly likely Blue Duck killed Will Green by mistake. He was so drunk he couldn't even catch your signals at the card table. With this trial, you see your way clear to get rid of Blue Duck, who has become a reckless partner. He drinks too much, and if during one of those drinking binges, he blurts out how you and he have been cheating at cards for years, you're through.

But then on the other hand, if you testify against him, Blue Duck will probably spill the beans about your cheating anyway—to a whole court full of people. You've got a dilemma on your hands. You can't shoot everybody in town in the back.

Life is so complicated.

Secret Information
Allan Pinkerton—Witness for the ?

As a detective, your job is to notice things. Listen to all the testimony and take notes. After the trial has been in session and all the witness have been called, you will offer yourself as a surprise witness for the appropriate side.

Secret Information
Belle Starr—Witness for the Defense

You don't care if Blue Duck killed Will Green or not. It's time to do what you do best, persuade people to do what you want. Since you are renown for tampering with juries, do your thing.

Blue Duck is a passing fancy, but he doesn't deserve to die. And besides, Will Green was getting on your nerves, wanting to buy your part of the cattle business. Perhaps you can snuggle up to that Cole Younger on the jury. He's a good looking guy.